Food Chains
The Unending Cycle

Margaret J. Anderson

with drawings by Gretchen Bracher

The Living World Series

ENSLOW PUBLISHERS, INC.

Bloy St. & Ramsey Ave.　　P.O. Box 38
Box 777　　　　　　　　　Aldershot
Hillside, N.J. 07205　　　Hants GU12 6BP
U.S.A.　　　　　　　　　　U.K.

> To Susan and Barry

Library of Congress Cataloging-in-Publication Data

Anderson, Margaret Jean, 1931-
    Food chains : the unending cycle / Margaret J . Anderson.
       p. cm.—(The living world series)
    Includes bibliographical references.
    Summary: Explores the concept of food chains and discusses their
importance in nutrient cycles and in the maintenance of ecological
balance.
    ISBN 0-89490-290-3
    1. Food chains (Ecology)—Juvenile literature. [1. Food chains
(Ecology) 2. Ecology.] I. Title. II. Series.
    QH541.14.A53   1991
    574.5'3—dc20   90-3282
                                         CIP
                                         AC

J 574.5 An (handwritten)

Printed in the United States of America

10 9 8 7 6 5 4 3 2 1

Illustration Credits:
Margaret J. Anderson, pp. 9, 26, 28 (bottom), 32, 38, 51; N.H. Anderson,
Entomology Department, Oregon State University, Corvallis, Ore., p. 15; John Barros,
p. 46; Gretchen Bracher, pp. 8, 11, 14, 16, 18, 20, 24, 44, 47, 52, 54; R.W.
Henderson, p. 58; John Hyde, Alaska Department of Fish and Game, p. 49; Paul
Komar, pp. 41, 42; Reproduced with permission, from *Exploring City Trees* by
Margaret J. Anderson © 1976 by McGraw-Hill, Inc., All rights reserved, pp. 23, 27,
28 (top); Reproduced with permission, from *Exploring the Insect World* by Margaret
J. Anderson © 1974 by McGraw-Hill, Inc., All rights reserved, p. 56; National Trust
for Scotland, p. 4; © Jane Thomas, pp. 6, 33, 34, 36.

Cover Photo: Don Alan Hall.

Contents

The seabirds nesting on the cliffs were the main food of the St. Kildan Islanders.

1 / The Chain of Eating and Being Eaten

The St. Kilda Islands lie about a hundred miles west of Scotland—four rocky, treeless islands jutting out of the Atlantic Ocean. They are the home of millions of seabirds and a few flocks of wild sheep. Stone walls and roofless cottages indicate that people once lived there too. For centuries, the islands supported a population of around two hundred people, whose staple food was the seabirds—gannets and fulmars—that nest on the thousand-foot high cliffs. The St. Kildans also tended fields of oats and barley, and ate wild plants, such as sorrel and dock and silverweed. Wool and sheepskins provided them with clothing, and peat or turf fueled their fires.

In the nineteenth century, contact with the British mainland introduced the St. Kildans to a wider world with a more varied diet and new ways of doing things. Many of the young people emigrated. This eventually led to the evacuation of St. Kilda in 1930. But even back in the days when the St. Kildans were self sufficient, their small islands were not really supplying all their needs. The gannets and fulmars that they ate foraged out in the Atlantic for fish. These fish had grown fat on zooplankton—small drifting animals in the sea—which had, in turn, dined on phytoplankton. Phytoplankton—drifting plants—capture radiant energy from the sun and minerals and gases from the seawater and convert them into food. So the St. Kildans were part of a food chain that stretched far beyond their own shores.

The island plants that the people ate also depend on the seabirds, which fertilize the land with their nutrient-rich guano droppings. The plant roots prevent the wind and rain from eroding the thin soil. And when a plant or animal dies, bacteria and other microbes go to work, sending the nutrients back into the food chain.

The Science of Ecology

Because the St. Kildans were dependent on birds for food, they learned the seabirds' seasons and their nesting places. They knew when each of the migrating species could be expected to return to the island and they knew which ones stayed there all year. They observed that gannets

Gannets with their young.

will lay a second egg if the first is taken. Fulmars, on the other hand, lay only one egg and don't lay again if the nest is disturbed. So the St. Kildans stayed away from the fulmar colonies on the cliffs at breeding time.

Although they didn't know the word, the St. Kildans were good practicing ecologists. Ecology is quite a recent concept in our vocabulary. The word was coined some 150 years ago from the Greek words *oikos* meaning "house" and *logos* meaning "study." Ecology is the study or science of the relationship between living organisms and their surroundings, including other animals and plants, the soil, and the climate. This makes it a complicated study because it combines botany, zoology, chemistry, physics, geology, and meteorology.

The word *ecology* is often misused and is sometimes given a negative slant. People talk about disturbing or destroying the ecology of an area, when they really mean destroying the habitat or environment. You can study ecology in a polluted lake just as well as in a clear mountain lake. The polluted lake, like the clear lake, is an ecosystem where living organisms are affected by, and have an effect on, their environment.

Food Chains and Pyramids

One main function of all organisms is to reproduce. This is true of whales, elephants, oak trees, people, dandelions, and mosquitoes, right down to the smallest microbe or virus. To do this takes energy and energy comes from food. So all organisms must eat. Plants are the producers at the beginning of the food chain. They contain a pigment called chlorophyll. In the presence of sunlight, water and carbon-dioxide are converted into carbohydrates.

Water + Carbon dioxide $\xrightarrow[\text{(Energy from the sun)}]{}$ Carbohydrates + Oxygen

This process is called *photosynthesis*, which means "putting together with light."

All animals are consumers. Some are primary consumers, or herbivores—getting their food from plants. Others are secondary consumers or carnivores. They get their food second-, or third-, or fourth-hand from plants by eating other animals. Then there are consumers, like ourselves, that are omnivores—eating both plants and animals.

As we go up the food chain from the producers to the primary and secondary consumers, the numbers of organisms decrease. If we counted the number of organisms at each step of the St. Kildan food chain, we would find that it took millions of phytoplankton to feed hundreds of thousands of zooplankton. These were eaten by thousands of fish, which in turn were eaten by the hundreds of seabirds that nourished just one person for a year. The numbers form a pyramid, with the millions of phytoplankton, or producers, at the base. This pyramid of numbers is just as true for the American eating hamburgers as for the St. Kildan eating birds. It takes 20 million alfalfa plants to feed four and a half calves, the equivalent of which can be consumed by one growing twelve-year-old in a year.

The pyramid is sometimes referred to as the Eltonian pyramid. It is named after Charles Elton, an ecologist, who spent one summer seventy years ago studying the eating habits of foxes on an arctic island. He counted all the small mammals and birds that were eaten by a fox, and the insects and worms that were eaten by the small animals. At first, he thought that the resulting pyramid of numbers was because a few big animals were the equivalent of many small animals. But when he calculated the weight, or biomass, at each level of the food chain, he found that this, too, formed a pyramid.

The Energy Pyramid

If we think of food in terms of calories instead of weight, we begin to see why it takes a large biomass of little animals to support a smaller biomass of big animals. A calorie is a measure of energy. It is the amount of heat energy required to raise the temperature of one gram of water by one degree centigrade. Plants convert radiant energy from the sun into food energy by photosynthesis. This energy is stored in their cells and is used as the plants breathe and grow and reproduce. Food energy is transferred into animals that eat the plants. But when stored energy is changed to another form, some of it gets lost. For example, if stored energy is changed into the energy of motion, a lot of energy is lost as heat. That's why you feel hot when you run fast.

Because organisms use up energy when they move and grow and respire and reproduce themselves, they can't pass on all the energy they consume in the form of food. The percentage of energy passed up the food chain is not, in fact, very high—only about 10 percent of the total energy at each level. The only energy coming into the system is the radiant energy of the sun that is captured by the plants, while at every step of the food chain, energy is lost. So we have another

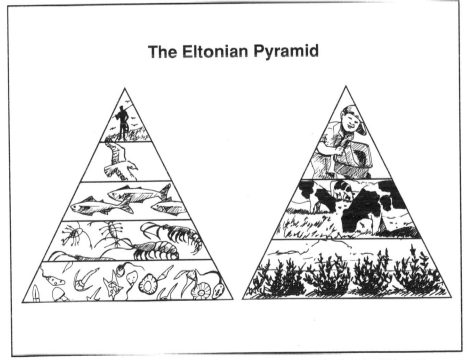

The Eltonian Pyramid

The number of organisms decreases with each step of the food chain. This holds true for a St. Kildan eating seabirds or an American eating hamburgers.

pyramid in the food chain—the pyramid of energy—which is also supported by plants at the bottom.

The flow of energy up the food chain is a one-way path with about 90 percent of the energy lost or used up at each step. The energy pyramid explains why there are fewer big animals in the world than small ones. You can't have more big predators than you have prey. You can't even have as many.

In addition to limiting the number of animals at the top of the food chain, the supply of energy limits the number of links in the chain. Starting with plants, there are usually no more than four or five links in the food chain. These links are known as trophic levels.

When stored energy is changed into the energy of motion, some energy is lost in the form of heat.

Web of Life

So far we have been looking at food chains as if they were a one-way street with no side roads. But it is more complicated than that. Meat-eaters obtain their food from a complex network of food chains. For example, the fox on an arctic island might eat a ptarmigan that lived on plants. Then it might catch a sandpiper that has eaten a spider that ate a caterpillar that ate a leaf. Or it could capture a seagull that

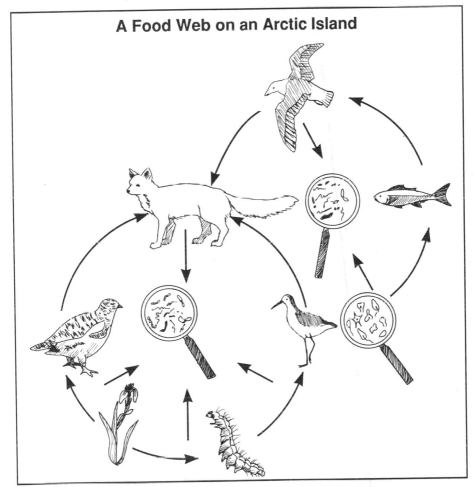

A Food Web on an Arctic Island

The food chain can be more accurately described as a web. The decomposers are important at all levels.

has dined on fish or shellfish that obtained their food from the plankton in the sea. It is more accurate to picture these pathways as a food web rather than as a food chain.

To understand the complexity of the food cycle we have to consider another important group of organisms that is connected with every step of the chain—the decomposers. Decomposers, such as bacteria and fungi, are at work everywhere, acting at every trophic level of the food chain. They break down wastes and dead bodies into raw materials, thereby making these materials again available to the food cycle. Without the decomposers the food chain would not be a one-way street. It would be a dead end.

The decomposers are so important that they require their own chapter. But to appreciate their role in recycling nutrients, we first need to take a look at the chemistry of the food cycle.

2/Nutrient Cycles

When we hear the word *recycle* we tend to think of bottles, aluminum cans, and old newspapers. Recycling, however, is as old as the world itself. All the chemical elements that are needed to support life are contained in the soil, the water, or the atmosphere. Unlike energy, there is no outside source. The same elements have to be used over and over again. Some of the calcium in your bones could have been present in the skeleton of a dinosaur that lived a hundred million years ago and in the shell of a sea creature that lived millions of years before that. Between times, it could have been part of a coral reef.

A cycle that involves both living organisms and the earth is known as a biogeochemical cycle, from *bio* meaning "living" and *geo* meaning "earth." Although you may not know it by that name, you are familiar with at least one important biogeochemical cycle—the water cycle. When rain falls from the clouds, it is taken up from the soil through the roots of plants or is drunk from rivers by animals. The plants and animals return some water directly to the atmosphere in the form of water vapor. Some may continue up the food chain and some is excreted in wastes to cycle through the earth and the rivers and the sea. Eventually that water evaporates to complete the cycle by once again becoming part of a cloud.

The Phosphorus Cycle

Phosphorus occurs in the rocks that make up the earth's crust. It is also an essential part of protein, and so it is found in all living organisms. When rain, wind, and ice gradually wear down rock, phosphorus is released, mostly in the form of phosphate (phosphorus linked with oxygen). Dissolved in rain water, this can then be absorbed by the roots of plants. It makes its way up the food chain when plants are eaten by animals. When organisms die, the phosphorus in their bodies is returned to the soil and the rivers through the work of the decomposers. The rivers carry the phosphorus to the ocean, where it is eventually deposited on the seafloor. After a long, long time, the phosphorus turns to rock. The phosphorous cycle is not measured in years or centuries but on a geological time scale.

Some of the phosphorus that washes into the sea is taken up by phytoplankton and becomes part of the ocean food chain. It may then return to the land in harvested fish or as guano from seabirds. Along

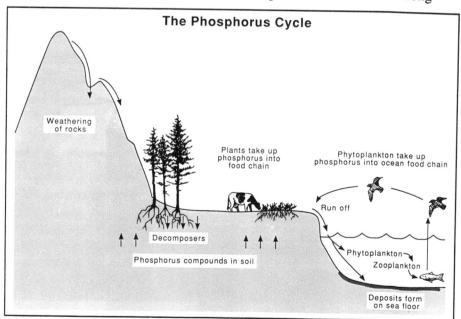

Phosphorus cycles through the earth, the sea, and living organisms.

the coast of Peru there are centuries-old deposits of guano from cormorants feeding on fish in the coastal waters. These guano deposits are important to Peru's economy because the phosphorus that they contain is in demand as an ingredient in fertilizers.

Applying phosphate fertilizer—either from guano or from phosphorus extracted from rock—results in higher yields of crops. Unfortunately, when fertilizer washes off farm land into rivers and lakes, it can create an environmental problem. Phosphate in water works the same way as it does on land—it promotes plant growth. Increased production on land is good, but increased plant production in the water systems is not so good (from our point of view).

A sudden increase in the amount of algae in a lake is called an algal bloom. Like land plants, water plants and algae manufacture their food by photosynthesis, giving off oxygen. They also use up oxygen in respiration. But it is when they die that the trouble begins. Dead

Phosphates in lakes promote the growth of water plants.

plant tissues sink to the bottom of the lake, where they are decomposed by bacteria. Oxygen from the surrounding water is used up in the process of decay. Fish, such as trout, which need oxygen-rich water, cannot survive in these low-oxygen conditions. Other fish, like carp, that require less oxygen, take their place. It is not poisonous wastes, but an overabundance of phosphate from fertilizers or from detergents that is often the main cause of lake pollution.

The Nitrogen Cycle

Long ago, farmers noticed that if they grew crops on the same ground year after year, yields declined. But if they gave the field a rest every third year or so, they got good harvests indefinitely. While the soil was resting, bacteria were converting proteins in dead organic material into nitrates. Other bacteria in the soil were taking nitrogen from the air and combining it with oxygen to form nitrates. The bacteria were, in fact, fertilizing the field by making nitrogen available to the plants.

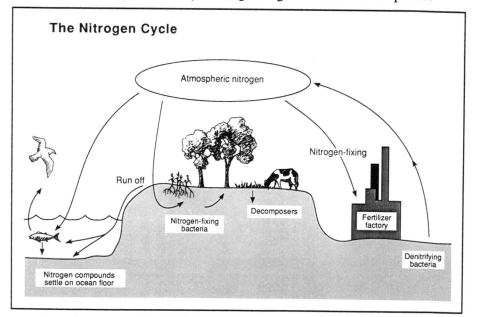

The Nitrogen Cycle

Atmospheric nitrogen

Nitrogen-fixing

Run off

Nitrogen-fixing bacteria

Decomposers

Fertilizer factory

Denitrifying bacteria

Nitrogen compounds settle on ocean floor

Bacteria are responsible for fixing nitrogen from the atmosphere. Another group of bacteria returns nitrogen to the air.

16

Nitrogen is the most common element in the atmosphere. The air we breathe is 78 percent nitrogen and 21 percent oxygen, leaving only 1 percent to be shared by other gases, including carbon dioxide. Nitrogen is an inert gas, which means it doesn't do very much. We breathe it in and we breathe it out again unchanged. It is, however, one of the elements that is necessary in the formation of protein.

Leguminous or pod-bearing plants, such as peas, beans, clover, or alfalfa, all have nodules on their roots that contain special nitrogen-fixing bacteria that trap nitrogen from the air. The bacteria living in the roots of a field of clover can add as much as 500 pounds of nitrogen per acre to the soil. Some of the nitrogen compounds stay in the soil, while some are used by the clover and continue up through the food chain—to the cow that eats the clover and then to the child that drinks the milk. Nitrogen compounds in animal wastes and dead organisms end up back in the soil, where they can be taken up by the plant's roots. In this way nitrogen can cycle through the food chain again and again without returning to the air.

In water, blue-green algae fix nitrogen, making it available to the ocean food chains. In the forest, bacteria associated with alder trees and some kinds of brush help to enrich the soil. Now we also have factories that imitate the work of the nitrogen fixers, extracting nitrogen from the air and turning it into nitrates for fertilizer.

So far, we have a one-way system, with the nitrogen-fixers taking nitrogen from the air, and everything else using it. If this were the whole story, the percentage of nitrogen in the air would gradually decrease. Since it remains the same, there must be a way of sending nitrogen in the other direction. The answer lies in yet another group of bacteria—the denitrifying bacteria.

Denitrifying bacteria are found in marshes and in the mud at the bottom of lakes and estuaries. They feed on the rich accumulations of water-logged plant material. One of the problems for organisms living in such a place is lack of oxygen, but these bacteria have solved that. They take oxygen from nitrates, releasing nitrogen gas into the atmosphere. The nitrogen cycle is dependent on a great many unseen bacterial helpers.

The Carbon Cycle

Almost 93 percent of our body weight is made up of three elements—hydrogen and oxygen (the two elements present in water) and carbon. Carbon accounts for over half the dry weight of living matter. Like phosphorus, carbon enters the food chain through plants, but very little comes from the soil. Plants absorb carbon in the form of carbon dioxide gas from the atmosphere and convert it into carbohydrates by the process of photosynthesis. Some carbon is returned to the atmosphere as carbon dioxide when the plant respires. Some ends up in the animal that eats the plant, where it may be transformed to fat or protein or to more complex carbohydrates. Some returns to the atmosphere

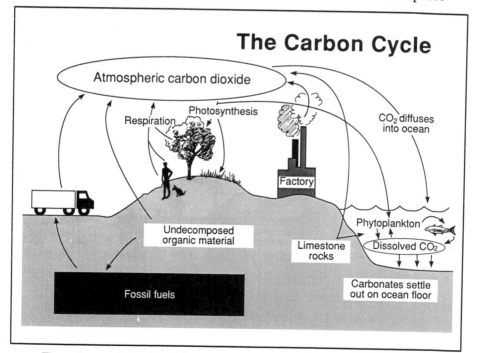

The Carbon Cycle

Atmospheric carbon dioxide

Respiration　Photosynthesis

CO_2 diffuses into ocean

Factory

Phytoplankton

Undecomposed organic material

Limestone rocks

Dissolved CO_2

Fossil fuels

Carbonates settle out on ocean floor

The carbon cycle involves living organisms and the earth, ocean, and atmosphere. Emissions from automobiles and factories are adding to the amount of carbon

when the animal breathes and some is excreted as waste. The carbon travels farther up the food chain if the first animal becomes the dinner of another animal. At some point, the decomposers go to work on the dead organism, returning the carbon to the atmosphere as carbon dioxide or storing it in the soil.

The length of time the carbon remains in an animal depends on what other elements it joins along the way. If, for example, a carbon atom links up with three oxygen atoms and a calcium atom to form calcium carbonate, it may become part of an animal's skeleton and be in that form for a long time, even after the animal is dead.

The oceans contain about 50 times more carbon dioxide than does the atmosphere. Some of the carbon in the ocean is absorbed directly from the atmosphere in the form of carbon dioxide, which is readily soluble in water. Carbon also washes off the land in the form of calcium carbonate when limestone cliffs are worn down by wind and rain. The limestone was formed on some ocean floor in the distant past when countless sea creatures died, leaving their shells behind. Limestone provides a kind of "savings account" of carbon in the *geo* part of the biogeochemical cycle.

The Greenhouse Effect

Although carbon dioxide is essential to life on earth, it only accounts for about 0.03 percent of the gases in the atmosphere. A complex balance has been reached between the amount in the air and the amount dissolved in the sea. The oceans act as a buffer zone and reservoir, keeping the level in the atmosphere fairly constant. But recently—during the last two centuries—burning fossil fuels has introduced a significant new pathway into the carbon cycle. Millions of years ago, a great many carbon atoms dropped out of the cycle when the remains of plants were trapped as coal or oil in sedimentary rock. They get back into the cycle again when we burn these fossil fuels. The annual rate of release of carbon dioxide from factories and automobiles is about 5 billion tons, which is equal to 10 percent of the amount used in photosynthesis. Although this is a great deal, the effect on the planet of this tremendous quantity of carbon dioxide is still hard to measure and to predict.

At first, it was thought (or hoped) that the oceans would absorb the surplus, but apparently the ocean is not keeping up. Measurements of carbon dioxide taken in Antarctica and on the top of a mountain in Hawaii show that the level of carbon dioxide in the air is rising, even in places that are far away from highways and factory chimneys.

It could be argued that an increase of carbon dioxide would help to make plants more productive. However, one of the characteristics of the carbon dioxide in the atmosphere is that it absorbs the heat escaping from the earth's surface and bounces it back to earth. More carbon dioxide results in more heat being bounced back. This has been called the "greenhouse effect" because the carbon dioxide is acting

Preserving forestland is important in the carbon dioxide balance between the earth and the atmosphere.

like greenhouse glass. Global warming trends could have long reaching effects. The ice caps might melt, raising the level of the sea and flooding coastal cities. Rainfall patterns might change throughout the world.

Another way we are upsetting the carbon balance in the atmosphere is by cutting down and burning forests, especially the tropical rainforests. The carbon that was tied up in all those trees is released. At the same time, we are losing these efficient plant factories that absorb carbon dioxide from the atmosphere. Over a long period of time, as the water in the ocean circulates, the extra carbon dioxide in the atmosphere will be absorbed. In the meantime, we should be enlisting the help of the forests by planting more trees than we cut down.

3/The Decomposers

Anyone who has had the chore of raking leaves in the fall knows that trees invest a great deal of energy in producing leaves, only to have them end up on the ground. Even evergreens lose their leaves, although they don't all fall at the same time. The organic matter that accumulates on a forest floor—leaves, needles, twigs, animal wastes, cast skins, dead beetles and so on—is known as detritus.

Detritus accounts for a lot of nutrients and energy. Only about 10 percent of plant material moves up to the next trophic level by being eaten, so the other 90 percent has to have some way of getting back into the food chain when it dies. Otherwise, all the nutrients in the world would soon be locked up in dead plant material. The carpet of leaves and needles would get deeper and deeper every year, eventually burying the trees.

The key to why this doesn't happen is found in a large group of organisms called decomposers. Their job is to break down the organic compounds in dead plant and animal materials, releasing the inorganic nutrients into the environment. These decomposers are so important that living organisms are sometimes divided into three major groups instead of two. The two groups that we are familiar with are the plants and animals—the producers and consumers in the food chain. The third group is the reducers or saprobes. The word *saprobes* comes from *sapros* which is Greek for "rotten." Saprobes are the bacteria, yeasts, and fungi that live on rotting plant and animal tissues. Most are too small to be seen with the naked eye. Fungi, however, are visible at the reproductive stage when mushrooms appear above ground.

Saprobes and Shredders

Green plants feed by photosynthesis, and have evolved leaves supported by stems to enable them to reach up and take advantage of the sunlight. Animals feed by eating plants or other animals. For getting around to look for food and for digesting it, animals need complex bodies and complex organs. But saprobes spend their entire lives surrounded by their food. They do not have to go looking for it and have no need for legs or fins. They eat by releasing enzymes that

Fallen leaves contain a lot of nutrients and energy.

dissolve the organic material around them, and then they absorb this predigested food. They can get along without stomachs for digesting food or large bodies for storing it.

Although fungi and bacteria are efficient at attacking plant tissues, they need help from a group of animals known as shredders. A scientist demonstrated that leaves on the forest floor took a long time to decompose if they were enclosed in fine-mesh bags, even though lots of bacteria were present. The mesh bags had excluded the shredders. Shredders chew the organic matter into smaller pieces, providing more surfaces for the saprobes to work on. The shredders also stir things up, continually recirculating both the detritus and the saprobes. Saprobes

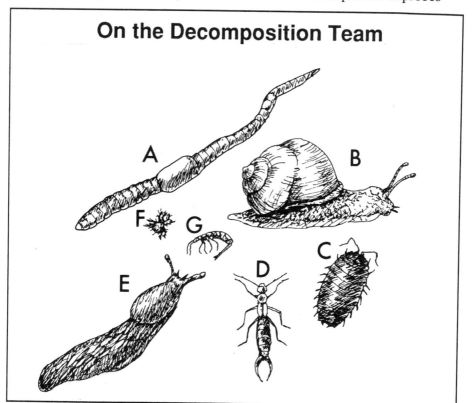

On the Decomposition Team

Shredders are important agents of decomposition: A) earthworm; B) snail; C) sowbug; D) earwig; E) slug; F) mite; G) springtail.

24

can't get around on their own, but they often hitch a ride on, or even inside, the shredders. Shredders include worms, earwigs, sowbugs, slugs, and snails. Some of these creatures are despised in the garden, but they are essential to the recycling business.

Worms play an especially important role in decomposition. During the day, they tunnel through the earth and detritus. At night, they come up to the surface to mate and to drag decaying leaves down into the soil. Their tunnels provide pathways through the detritus layer for oxygen, which the saprobes need for respiration.

Other inhabitants of the detritus world include algae, slime molds, and a variety of insects, mites, and crustaceans. Some of these animals are picky eaters, requiring certain organic acids and vitamins to survive. Others eat anything that comes their way. The analysis of the gut of a small insect called a springtail turned up plants, fungi, bits of decaying earthworm, another springtail, and fecal material. Wastes from one organism provide nutrients for another. The shredders often eat saprobes along with the partially digested food that surrounds them. This makes it hard to separate out the trophic levels. In the first chapter we discovered that the food chain can be more accurately described as a web. At the decomposer level, the web becomes tangled. Worms get much of their nourishment from the saprobes and the food they were breaking down, and the saprobes digest the feces and dead bodies of worms.

A Compost Heap

In a compost heap, plant wastes are recycled at high speed. Leaves, grass clippings, coffee grounds, potato peelings, and wilted salad create exactly the right environment for the decomposers. The bacteria multiply rapidly and the process of decay speeds up. With the saprobes hard at work, energy, in the form of heat, is released. If you're not squeamish, stick your hand into a compost heap. You'll find a hot, humid world below the surface of the grass clippings. Temperatures can reach about 55°C (130°F).

If you look at compost closely, you should be able to make out networks of fungal threads as fine as spiderwebs. Spread a little in a container. With the aid of a hand lens and a bright light, you'll see a myriad of creatures: eight-legged mites scooting about at high speed— far too fast for you to count their legs, springtails hopping like jumping jacks, and nematode worms wiggling across the moist surface of a rotting leaf. A millipede may crash into view like a dinosaur invading your miniature world. Or is it a centipede? They are easy to tell apart—though not by counting all their legs as their names suggest. You need only count the legs on any one segment of their bodies. Centipedes have one pair of legs per segment, millipedes have two pairs.

The earwig, with pincers at its rear end, is another easy-to-recognize "giant" that is found around the compost heap. Sowbugs also

Leaves, grass clippings, and vegetable wastes all decompose at high speed in a compost heap.

gather on the fringes. They are crustaceans, with gill-like breathing organs that need to be kept moist. Some, nicknamed pill bugs, roll up for protection. Whether living in a compost heap or hiding under a rotting board, they are part of the team of creatures that return dead organic material to the soil.

Fallen Trees

In contrast to the compost heap, a fallen tree is an example of decomposition in slow motion. The same processes are taking place, but it can take centuries for all the nutrients from a decaying log to get back into the soil.

The first creatures to go to work when a tree crashes to the forest floor are usually wood-boring beetles, carpenter ants, termites, and mites. The wood-boring beetles may have been living in the tree while it was still standing. Trees that are blown down in a storm are often those that were already weakened by insects and fungi.

A fallen tree is an example of decomposition in slow motion

27

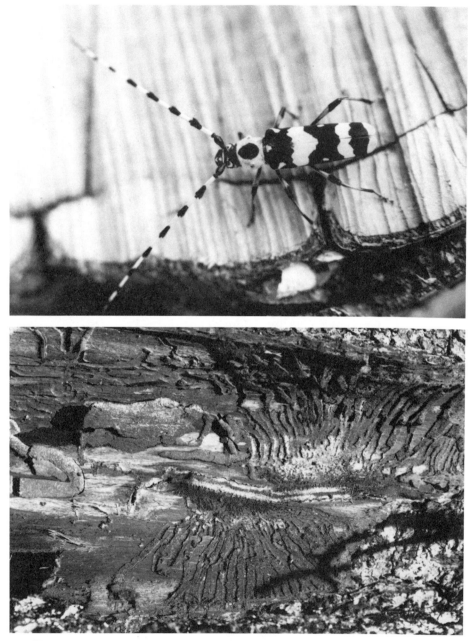

The wood boring beetle (top) has larvae that tunnel into wood allowing fungi and moisture to penetrate. Bark beetle larvae create a pattern of galleries under the bark (bottom).

Bark beetles lay their eggs in either live trees or trees that have just fallen. The female beetle chews a tunnel about 24 inches (60 cm) long with grooves off to either side, where she lays her eggs. When the eggs hatch, the larvae dine on the green tissue of the inner bark. If you peel the bark from a fallen tree, you can often find bark beetle galleries and tunnels—a hidden work of art created by beetle larvae as they eat their way through life.

Termites have microscopic animals, called protozoa, in their guts that assist in the breakdown of the tough lignin in the cell walls of wood. We are accustomed to think of these insects as destructive enemies—and they are when they eat our buildings. But they are also a necessary part of the forest food chain. By chewing the fallen tree into smaller pieces, they make the wood available to other insects and mites and saprobes. They let air penetrate the rotting wood and provide transportation for bacteria and fungi.

A rotting log can act like a sponge, holding more water than the surrounding soil. Plants and seedling trees, taking advantage of the trapped moisture, sprout along its length. The log is a nursery that will help raise the next generation in the forest.

We talk of trees as being a "renewable resource" because we can replant them when we cut them down and have a new forest in forty years or so. However, the trees removed from the forest for lumber, wood pulp, or firewood contain a lot of nutrients. There's a limit to the number of generations of trees that can be grown if no nutrients are returned into the cycle and if the soil is allowed to erode.

Lakes and the Sea

Detritus is an important source of food in water as well as on land. The weight of dead material at the bottom of a lake is usually several times the weight of the living organisms that the lake supports. Much of the detritus in fresh water—in springs, streams, and lakes—comes from the vegetation along the banks and shore line. Less is derived from plants and animals living in the water. Most of the detritus in the ocean comes from the remains of the sea plants and animals that lived there, particularly from plankton.

Decomposition takes place very slowly at the bottom of deep water due to the scarcity of oxygen and low temperatures. However, the stirring effects of wind and currents produced by changes in temperature carry some oxygen down to the detritus layer and bring nutrients to the surface, speeding up the process.

In some aquatic environments, such as marshes, oxygen is completely lacking. In those places, anaerobic bacteria and yeasts help to break down the detritus. *Anaerobic* means "able to work in the absence of oxygen." Anaerobic bacteria, which include the denitrifying bacteria mentioned in the last chapter, release nitrogen, phosphorus, or sulfur into the environment. The sulfur accounts for the "rotten egg" smell you sometimes notice around marshes or polluted rivers. Polluted rivers often contain large amounts of dead matter and little oxygen due to algal blooms.

Anaerobic bacteria are not nearly as efficient as bacteria that use oxygen. This, however, has been to our advantage. Anaerobic bacteria were unable to completely decompose the vast amounts of dead organic matter in the swamps and marshes of the distant past. This became the peat and coal that is now under the earth's surface. If the decomposers had done a better job 300 million years ago, we would not have coal to burn today.

4/Food Chains on Land

Before everyone left St. Kilda, two species of mice lived on the islands—the field mouse and the house mouse. Field mice nested among the rocks and ate seeds and grass. House mice lived in the cottages, eating whatever they could find. Both kinds of mice raided the St. Kildan's storage buildings and made their homes in the walls around the village. After St. Kilda was evacuated, the house mice became extinct within a few years. Even though their food and habits had overlapped with those of the field mice, they couldn't compete successfully in field-mouse territory. Two very similar species cannot share the same food and home. Each species requires its own niche. *Niche* in this sense means more than a place to live. It describes an animal's—or a plant's—whole lifestyle. A niche is rather like a job description.

A single oak tree provides food and shelter for a great many different kinds of animals from squirrels to caterpillars. Although they all apparently share the same address, each creature has its own niche. Squirrels scamper up the trunk and along the branches and eat the acorns. Robins roost on outer branches and eat insects. Caterpillars munch on leaves, aphids eat sap, and ladybugs eat the aphids.

31

Some animals on an oak tree lead a very restricted life. The larvae of a tiny wasp grow up inside strange growths on the leaves called galls. Galls are also known as "oak apples" because they look like small speckled apples. Another species of wasp is parasitic on the first species. It lays its eggs in a growing larva inside the gall. When this happens, the wasp that finally emerges is not the one that caused the gall to form. In seasons when the parasite is common, very few of the gall-forming wasps emerge as adults. The following year, there are fewer galls. The parasites then have a hard time finding places to lay, and so their numbers decline.

The availability of food and of niches imposes a check on the size

Galls on oak leaves, also called oak apples, are the home of the larvae of a small wasp.

of the population of a particular species, while the variety of niches allows a great number of species to share the tree. This variety of living forms in a habitat is known as biodiversity. The biodiversity in a single oak tree results in a complex network of pathways in the food web, but it doesn't begin to compare with the complexity of the food webs in trees in the tropics.

The Tropical Rainforest

In the tropical rainforest, it is always summer and heavy showers fall nearly every day. Viewed from a river or from the edge of a clearing, the forest is the dense, tangled jungle of Tarzan movies. However, in the heart of the rainforest the floor is relatively open—a dim, damp place with green light filtering down from the distant tree tops. Huge buttressed trunks support the high, leafy ceiling, which is known as

Hundreds of different kinds of trees grow in the tropical rainforest.

the canopy. Trees compete with one another to secure a share of the sunlight.

Less than ten different species account for the trees in an old-growth coniferous forest in the Pacific Northwest, and up to forty species can occur in a temperate deciduous forest. In the tropical rainforest, however, three hundred species of trees can be found in an area of two square kilometers. These trees provide niches for a myriad of creatures. One entomologist collected twelve hundred different kinds of beetles from a single tree.

Most of the action in the rainforest is up in the brightly lit canopy. Birds with strangely shaped beaks, some adapted for cracking nuts, others for searching out insects from under bark, flit from one tree to the next. Huge butterflies float among the blossoms. Snakes and lizards and monkeys are as at home in the treetops as the birds. In the Amazon forests, monkeys, anteaters, and porcupines perform

Bright colored birds nest in the treetops.

acrobatic tricks with their prehensile tails—a sort of fifth leg that helps them hang on in their arboreal world.

Some creatures are brightly colored; others are well camouflaged. Hiding is a way of life in the canopy, where predators abound. The sloth, with green algae growing in its fur, looks more like a clump of moss than an animal as it hangs upside down from a branch. Snakes mimic vines, butterflies resemble flowers and leaves, and the stick insect looks exactly like a lifeless twig until it decides to walk away.

The trees provide a home for other producers as well as for consumers. Plants perch in the upper branches of tall trees, so that they, too, can share the sunlight without the effort of growing long stems. Perching plants, such as orchids and bromeliads, are called *epiphytes*. They take no nutrients from the tree and none from the soil. How then do they obtain the raw materials they need for growth?

The bromeliad, a member of the pineapple family, has solved the problem by having its own private food and water supply. The plant develops from a seed that has lodged in a crevice on a tree branch. As it grows, its overlapping leaves form a cup where rain water gathers. Soon mosquito larvae are sporting about in this miniature swimming hole. They live on detritus that falls into the water. Then a dragonfly may drop by and lay some eggs. When the young hatch, they grow fat on mosquito larvae. Snails feed on green algae growing on the damp walls of the little pool. A poison dart frog carries a tadpole all the way up from the damp forest floor, where it hatched from an egg, and deposits it in the water. Animal wastes and rain rich with nutrients washed from the leaves overhead accumulate in the pool. The bromeliad has fine hairs that reach into the nutrient-rich water within its leaves and absorb food. As the plant grows and blooms, insects are attracted to the blossoms and nectar. Birds and lizards snap up the insects.

The bromeliad pool is another example of a detritus-based food chain. The primary producer is not the tree, nor is it the epiphyte. The epiphyte is getting its nutrients from the dead organisms and wastes within the cup formed by its leaves.

In the rainforest, detritus doesn't have the chance to accumulate under the trees the way it does in the temperate forest. Fallen leaves and twigs are snapped up by ants. Anything they miss decomposes quickly in the hot, wet climate. Nutrients are immediately absorbed by the roots of the trees. The nutrients cycling through the tropical rainforest are nearly all in the living organisms and not in the soil.

Until recently, the tropical rainforests were fairly inaccessible. The humid climate, wild animals, and tropical diseases kept most people away. Within the last decade, however, modern road-building machinery, medical advances, and the pressures of population growth have turned the tropical rainforest into a new frontier, threatening its existence. People are cutting and burning the trees to make way for agricultural land that will have almost no biodiversity.

The rainforest, the home of an amazing variety of creatures, is in danger of destruction.

In Brazil, corn, beans, and sorghum are being planted to feed the ever-growing population. Large tracts of forest are also cleared for grazing beef cattle to provide cheap hamburgers for people in the United States. This cheap hamburger comes with high hidden costs in energy. As well as the energy lost between one trophic level and the next—turning grass into fat cattle—fuel energy is required to fly the meat all the way from Brazil to markets in North America.

Higher than the energy cost, however, is the cost to the land itself. In spite of the tall trees it supports, the red soil is not fertile. Part of the environmental tragedy of the tropics is that the soil isn't suited to growing annual crops. After a year or two, the harvests fail and the fields are abandoned. But the forest cannot easily re-establish itself on the impoverished soil.

At the present rate of destruction, the tropical rainforest, with all its amazing variety of species, will be gone in just 25 years. So little is known about the many different kinds of plants and animals in these forests that we don't even know what we're losing. Many of our medicines originally came from the tropics. More may be there to be discovered. We are just beginning to find out how many organisms share the rainforest and now we are on the verge of losing it.

Grasslands

Grass is one of the most common and thriving plants in the world. It requires far less water than trees do. Its roots form a dense mat that takes up water quickly when it rains and holds the soil in place during dry periods. An unusual feature of grass is that the leaves grow from the base, whereas the leaves of other plants grow from their tips. Grass can thus withstand repeated grazing—either by a cow or a lawn mower. Grazers actually help maintain grassland by keeping other plants from getting established.

Grass is the staple food for a variety of animals, including ourselves. All our cereals—wheat, oats, barley, rice, and corn—are in the grass family. We eat the seeds and leave the rest of the plant—

which is hard to digest—to animals with stronger stomachs. Mammals that live on grass leaves have a special compartment in their gut containing bacteria that help them digest the tough cellulose walls.

A large area of North America used to be grassland or prairie, supporting a wide variety of herbivores—insects, mice, prairie chickens, antelopes, and bison. These provided food for hawks, owls, coyotes, wolves, and people. Within the last hundred years most of the native grasses have been replaced by high-yield varieties of wheat and corn. This is an efficient way to produce food, but it also provides an unlimited food supply for certain rodents, insects, and bacteria. Monoculture cultivated plants—whether in a Douglas-fir forest or in a strawberry field—need to be protected from outbreaks of pests and disease. The search for new varieties of plants that are disease resistant and for safe chemicals to combat insects and disease is never ending.

Grazers help to maintain grassland by keeping other plants from getting established.

The Desert

About 14 percent of the earth's land surface receives less than 10 inches (25 cm) of rainfall a year. Lack of water cuts down on primary production. Plants that grow in the desert are adapted to withstand drought. They may have small leaves, or even no leaves at all. Cactus leaves are reduced to needles, and the plants manufacture food and store water in their stems.

The tough, thorny plants of the desert support consumers that have also learned to conserve water. The kangaroo rat—which is related to the mouse, and is neither a kangaroo nor a rat—lives entirely on dry seeds. It never drinks a drop of water, but releases water from its food. It has super-efficient kidneys, no sweat glands, and a long winding nasal passage so that no water is wasted when it breathes. Like many of the desert animals, it escapes the heat of the day by burrowing into the sand and coming out to feed at dusk.

Desert conditions can sometimes result from overgrazing. The introduction of deer and oppossums into New Zealand, where there were no predators to hold them in check, stripped away the vegetation. Soil was then lost through erosion, leaving wide areas of unproductive land. This points out the importance of predators farther up the food chain. The producers at the bottom of the chain suffer if the predators don't keep the herbivores in check. Even the land is part of this complex system and is affected by the plants and animals that use it.

5/Food Chains in the Sea

The ocean would seem to be the ideal environment for photosynthesis. There is no shortage of water, the concentration of carbon dioxide is higher than it is in the atmosphere, and rivers keep adding to the supply of minerals and nutrients. The temperature, from 0°C to about 30°C (32° to 86°F), is a suitable range for plant life. In spite of this, the ocean is not nearly as productive as the land.

One factor that restricts plant life in the ocean is lack of light. Only a narrow surface layer about 350 feet (100 meters) deep, which is known as the photic zone, receives enough light to allow plants to manufacture food. Nutrients, however, tend to accumulate on the ocean floor, making them inaccessible to plants near the surface. The problem of the separation of the nutrients and the photic zone is more severe in deep than in shallow water.

Another difficulty sea plants encounter is the restless nature of the ocean with its waves, tides, and currents. Because of this, the most successful plants are phytoplankton—tiny, free-floating unicellular plants that survive by literally going with the flow. These primary producers of the ocean are here today and gone tomorrow. They are snapped up by the zooplankton almost as fast as they reproduce themselves. There is no permanent crop of producers in the ocean. The closest equivalent to a forest or a prairie is found in the beds of giant kelp that grow near some shores. The brown fronds of these large seaweeds are kept in place by holdfasts attached to the rocks. Seaweeds, like land plants, contain chlorophyll and manufacture carbohydrates by photosynthesis. Although brown and red pigments often mask the color of the green chlorophyll, it is still there. The holdfasts are not, however, the same as the roots of land plants because they do no absorb nutrients.

The Intertidal Zone

The zone between the high-tide and low-tide marks offers a great assortment of habitats, so a wide range of plant and animal species make their homes and find their niches there. Sea anemones, shellfish, and crabs feast on the phytoplankton, zooplankton, and detritus

Plants live only in the surface layer of the ocean, where light can penetrate.

41

brought in by incoming tides. The breaking waves are rich in oxygen. Yet those same waves can be hazardous to plant and animal life. The creatures that live in the intertidal zone have to deal with being submerged in the sea as well as being exposed to the air. Some animals, such as sea anemones, sea urchins, oysters, and starfish live near the low tide mark and are under water most of the time. Others, especially those with shells like limpets and mussels, can stand longer exposure to the air. Closer to the high tide line, you find barnacles and snails.

The constant shifting of sand makes the beach a more difficult environment for animals than a rocky shore. There is, however, more life on a flat, empty beach than meets the eye. If you watch a flock of sanderlings close to the breaking waves, you'll see that they are constantly jabbing the wet sand with their beaks in search of crustaceans, worms, and clams. These burrowing creatures live on the tiny one-celled plants and animals that exist in the water trapped between the sand grains. Crustaceans—tiny shrimps and crabs and copepods—have succeeded in the ocean the way insects have succeeded on land.

Breaking waves are rich in oxygen.

In the warm waters along tropical coasts, colonies of small animals related to jellyfish and sea anemones form coral reefs. Coral reefs are the rainforests of the ocean in terms of biodiversity and production. The permanent part of the reef is formed from the calcium carbonate in the coral animals' skeletons. Algae live inside the skeleton as well, contributing food by photosynthesis. Other kinds of algae add calcium carbonate to the framework of the reef. The pounding waves enrich the water with oxygen and bring in a continual supply of nutrients. Colorful and bizarre worms, crabs, shellfish, and snails find food and hiding places in the reef. Luminous fish glide and dart through the clear water like underwater butterflies and birds.

Because there are so many different creatures living in a coral reef, the food webs are very complex. The Great Barrier Reef, off the coast of Australia, is the home of a giant starfish called the crown-of-thorns that lives on the coral animals. At the same time, the coral feeds on young starfish. Another predator, a giant snail, eats the crown-of-thorn adults. The giant snail population dropped recently, allowing the crown-of-thorns population to increase. Starfish began eating up all the coral animals. With fewer coral animals eating crown-of-thorn babies, more starfish grew to be adults with an appetite for more coral animals, resulting in the death of parts of the reef. The dying reef, which provided homes for a large community of creatures, is being eroded and broken up by the waves. Thus the disappearance of a predator high on the food chain can have an effect that goes beyond its immediate prey.

The Open Sea

Some areas of the open sea are more productive than others. The waters of the North Sea, for example, are very good for fishing. We can often explain why land plants grow well in some places and not in others by looking at the amount of rain in different areas. But why should some parts of the sea be fertile and others be desert? Obviously not because of a shortage of water!

A Food Web in the Ocean

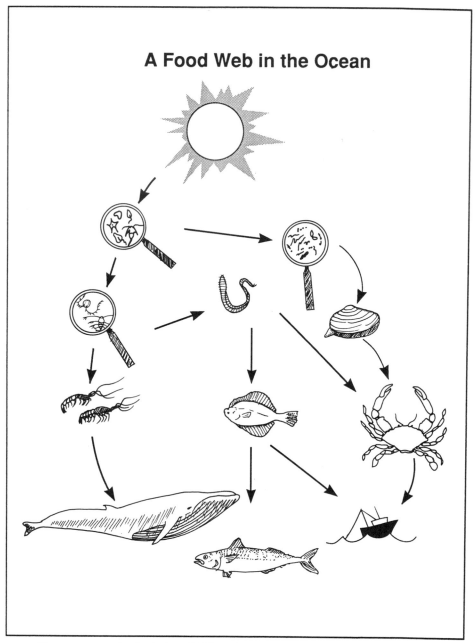

Crustaceans are the food of small fish, which are eaten by bigger fish. Some whales dine on krill, skipping the intermediate steps.

The explanation is found in the way the ocean currents distribute the nutrients. When the currents bring nutrients to the surface, the phytoplankton grow well. Scientists talk of a "phytoplankton bloom" even though these tiny plants do not flower. The rich areas of the ocean are not chemically different from the rest. They are just more stirred up, with nutrients circulating in the photic layer where the phytoplankton can use them. Upwelling currents in the North Sea have made it a rich fishing ground. Long-term changes in these currents could have devastating effects on the fish population.

Most of the food chains in the open sea involve little fish eating crustaceans and then being eaten by bigger fish. These bigger fish are eaten by still bigger fish. Some large marine mammals, like the baleen whales, skip the intermediate steps in the food chain and dine directly on the plankton. The blue whale, the biggest animal that has ever lived, eats krill, a sort of alphabet soup of tiny crustaceans. This gets around the loss of energy that occurs at each trophic level of the food chain. Even so, the blue whale has to devour millions of creatures to get sufficient energy for its huge body.

Although plants cannot live below the photic layer, the dark deep provides a home for a wide variety of animals, many of them like creatures from a science fiction movie. They are all detritus eaters or predators, directly or indirectly dependent on a rain of dead bodies from above. Strange worms and starfish scavenge on the ocean floor. Some fish have lost the use of their eyes. Others have developed light-producing appendages which attract their prey. Their cousins swimming in the surface layer are blue, green, and silver, but the bottom dwellers are mostly dark red or black.

Chemosynthesis

In 1977, a team of scientists located some hot spots on the floor of the Pacific Ocean in the Galapagos Rift. Water that had seeped down into the hot layer under the earth's crust was being forced back up through vents, creating hot springs on the ocean floor. When the scientists developed photographs—taken with underwater cameras—of the area of these springs, they were amazed to find pictures of hundreds of giant clams. The clams were as big as dinner plates.

Water samples gave off the distinctive rotten-egg smell of hydrogen sulfide. The scientists had discovered a unique food chain that was based not on the radiant energy of the sun, but on chemical energy from inside the earth itself. Energy derived from hydrogen sulfide by the process of chemosynthesis (putting together with chemical energy) was enabling certain types of bacteria to grow and multiply in total darkness. These bacteria, and organisms feeding on them, provided food for bigger creatures, including the giant clams. The scientists discovered other vents, where crabs and starfish were thriving. A vast clam graveyard indicated where a vent had shut down. Without the hydrogen sulfide to keep the bacteria—the primary

These clams and crustaceans live in total darkness on the seafloor, and depend on chemical energy from inside the earth.

producers in this system—going, all the creatures that depended on them died.

People and the Sea

Fish is the only major source of food that is still hunted from natural populations. At present, fish accounts for only 10 to 20 percent of the protein in the human diet, but it is becoming an increasingly important source of food for our ever-growing world population. We have to be careful that the increase in demand does not result in the destruction of the fishing industry. Regulating fishing is difficult because nobody owns the sea, and rules require the cooperation and agreement of nations as well as of individuals. As the demand for fish increases, fishing becomes more profitable. More people take it up on a commercial scale. When fish become scarce due to over-fishing, the price of fish rises, and so fishing remains profitable beyond the time when

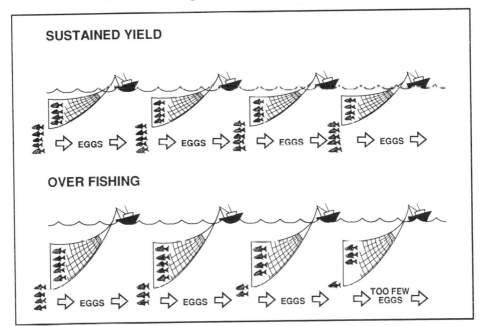

Overfishing occurs when not enough fish are left to produce a new "crop" of fish.

conservation of the remaining fish is possible. This scenario has already happened in the herring fisheries off the coast of Scotland.

If a farmer wants to stay in business, he doesn't kill off his entire herd when beef prices are high. He keeps enough prime breeding stock to make sure that he has a healthy herd the following year. The same principle obviously holds true for fish in the sea. However, because wild populations are more subject to the catastrophes of nature, a range of age classes of fish needs to be protected. This principle was discovered too late to save the anchovy fishing industry in Peru. Around 1972, the patterns of the currents in the Pacific Ocean changed, as they do from time to time. The resulting warmer water affected the fertility of young anchovies. In earlier years, when the anchovy population suffered from these temperature changes, older breeding fish survived, and reestablished the population. This time, the bigger fish had all been harvested. The anchovies have not come back with the return of cooler water, and this has affected animals all the way up the food chain.

It has been suggested that krill represents an enormous resource of food in the ocean. By eating low on the food chain, as many whales do, humans would avoid the loss of energy from one trophic level to the next. The trouble is that krill isn't very tasty. It could be converted into fish meal to feed chickens, but then we lose the energy advantage. We also have to take into account the fuel energy that would be required by the ships harvesting and transporting it. Another consideration is the effect that removing large amounts of krill from the food web would have on big fish and especially whales.

Aquaculture offers a more promising way of harvesting krill and small fish. Young salmon are reared in fish hatcheries and released as juveniles to forage in the ocean. When it is time for them to breed, they return to the hatchery, where they can be caught for food. In this way, we are harvesting krill and small fish from far out in the ocean without the effort of going out after them ourselves.

In addition to over fishing, another serious impact on the ocean food chain results from spilled oil. Big spills (like the one associated with the *Exxon Valdez* running aground on a reef off Alaska in 1989 and losing 10 million gallons of crude oil) make headlines for a month or two. People argue about the clean-up procedures and where to place the blame. Meantime, diving birds mistake calm oil-covered water for areas rich in food and dive through the surface oil. The oil clogs their plumage, so that they cannot fly. They also cannot regulate their temperature, and as they preen their feathers to remove the oil, some is swallowed. The birds die. Their bodies become part of the sediments, where the oil is ingested by worms, clams, oysters, and crabs. Long after the last of the clean-up crew has gone, the oil is still working its way destructively through the food chain.

Birds were some of the immediate casualties of the *Exxon Valdez* disaster in Alaska. The effects on the sediment layer will last from many years.

6/People and the Food Chain

Most of us are omnivores. We eat primary producers (plants) and primary and secondary consumers (animals). Our place in the food chain is right up there with the top predators, even though we don't have the right qualifications for the job. We don't have the teeth of a wolf, or the speed of a leopard, or the claws of a bear. Nor do we have the strength of a lion or the sight of a hawk. What we do have is intelligence. Our brains have brought us out on top.

During the last ten thousand years, we have figured out ways to change our staple food—grain—from a slender wild grass to a heavy-headed, high-yield plant. We keep our prey handy in fields or fatten them in feedlots. We have invented tools that kill animals and ways to preserve their meat. We no longer have to search out fruits and nuts in the forest. We grow them in orchards and ship them all around the world.

Brain power has allowed us to go a long way toward dominating nature. Intelligence enabled the St. Kildans to live in a place that would otherwise have excluded them. They survived on their remote islands because long-ago people had invented ropes and snares. Even more important, someone had figured out how to make a stone-age drying kiln so that stored food could last through the winter. Warm clothing, made from the furs of animals, allowed Eskimos to inhabit the Arctic. Four thousand years ago the people of northern Europe travelled over the surface of the snow on skis not much different from those we use today.

Our intelligence has allowed us to "go forth and rule the world." Our ability, particularly in the last 100 years, to modify the landscape to meet our own ends is staggering. Now we must prove that we have the intelligence and courage to cope with the problems we are inadvertently creating.

Agriculture

In some ways, ecosystems are like very complex machines with lots of interlocking, moving parts. The more complicated a machine is, the

Modern farm practices have greatly increased food production.

more likely it is to break down. But this doesn't hold for ecosystems. Complex ecosystems keep on working. This is another way of saying that biodiversity leads to stability. The rainforest, with its countless plants and animals all tied to one another in a network of relationships, doesn't change much if it is left to itself. When a huge tree, groaning under the weight of epiphytes and vines, crashes to the earth in a storm, seedling trees on the forest floor respond to the sudden increase in light. They grow quickly, and the gap is soon filled. The old tree will probably be replaced by a tree of a different species, but over the whole forest, the losses and replacements balance out. Outbreaks of pests or disease are rare because the large mixture of trees means that any one tree is isolated from other trees of the same species. All the different kinds of trees form a barrier against the spread of infection.

If bacterial root rot attacks a tree in a simpler ecosystem, such as a planted Douglas-fir forest, there's nothing to keep it from spreading

Trees growing in monoculture are susceptible to insect damage and disease.

to the next tree, and the insects that eat Douglas fir don't need to go far to find a suitable place to lay their eggs. When the eggs hatch, an abundance of food awaits the larvae. This also happens in a planted field of corn or potatoes or cabbages. With so much food available, conditions are right for a population explosion of the primary consumers—which are often insects and bacteria and not people. We call a one-species ecosystem a monoculture. Another problem that occurs in monocultures is that other primary producers—weeds—are eager to turn them into polycultures (planting many species).

Over the years, people have applied their intelligence to the challenge of maintaining monocultures. They found that ploughing and hoeing discouraged weeds. More recently, they have utilized chemical weedkillers. A whole army of chemicals—including insecticides and fungicides—has been developed to fight the war against pests. Our intelligence has revolutionized agriculture. In terms of feeding people, the wheatlands are far more productive than the prairies were. They have to be. There are far, far more people in the world today than there were one hundred years ago. But this increased production is not without its price. Serious problems can arise when chemical pesticides work their way up the food chain.

The Lesson From DDT

In the middle of this century, chemists discovered how to make organic compounds that kill living organisms. These new substances, called chlorinated hydrocarbons, don't occur in nature. The recycling bacteria cannot readily break them down. One of the best known of these chemicals is DDT.

Scientists assumed that the very small quantities of DDT required to kill an insect couldn't possibly harm people. DDT provided a quick solution to many insect problems, from cabbage caterpillars to malaria-causing mosquitoes. It was widely used, with positive results—such as the virtual elimination of malaria. Millions of lives

DDT Moves up the Food Chain

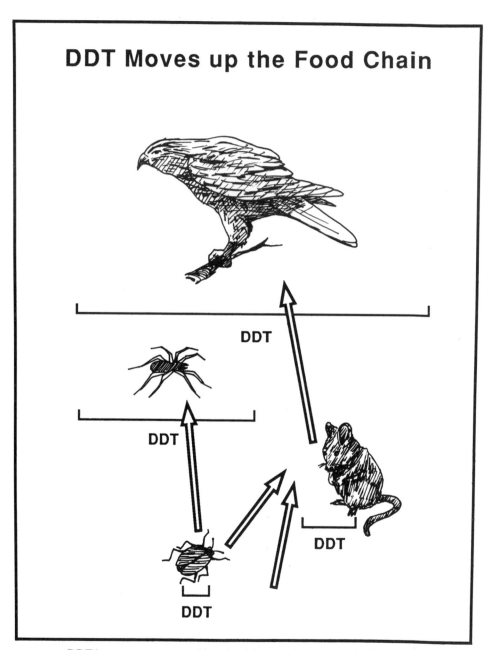

DDT becomes concentrated in animal tissue as it moves up the food chain.

were saved by the advent of DDT. But then birds began to die. And fish. And DDT was detected in cow's milk and chicken eggs.

DDT was working its way up through the food chain. When robins ate dead and dying caterpillars and fish ate mosquito larvae, they couldn't break down the DDT. They stored it in their body fat and tissues. Finally the stored DDT amounted to a lethal dose. When the animal died, bacteria couldn't break down the DDT either. DDT is a persistent poison that remains in the food chain. It has been detected in the fat of penguins living in the Antarctic, half a world away from where the poison was originally used.

Another problem with DDT, common to other insecticides, is that some insects are unaffected by it. These resistant insects produce resistant offspring. DDT no longer controls them, and so scientists have to come up with another poison.

Insecticides often kill predators as well as the plant eaters. Sometimes they even kill the predators instead of the prey. The birch leaf miner eats the tissue of birch leaves, spending its life between the upper and lower surface of the leaf. A small parasitic wasp lays its eggs inside leaf miner larvae, and thus keeps the leaf miner population in check. Chemical sprays killed off the wasps, but didn't harm the leaf miners inside the leaves. This resulted in such a severe leaf miner outbreak that many birch trees died.

A recent approach to the control of insect damage is the use of selective insecticides that kill only plant-eating pests. Predators can then help to keep populations of pests from building up again. Some plant-eating insects are also on our side. The caterpillar of the cinnabar moth eats tansy ragwort leaves, a weed that is poisonous to cattle. A small flea beetle eats tansy roots. By rearing and releasing cinnabar moths and flea beetles, scientists at Oregon State University have stopped the spread of the weed in pasture land. Promoting the consumers that are on our side is called biological control. The success of biological control depends on an understanding of the ecosystem.

Cinnabar moth caterpillars help to control outbreaks of tansy ragwort, a weed that is poisonous to cattle.

Subtraction and Addition

In an unmanaged forest, when a tree dies and falls to the ground, the nutrients are recycled into soil by the work of the decomposers. Other trees draw on these nutrients. But when a tree is felled and turned into lumber, the nutrients end up as part of a building a long way from the forest. In the same way, when a farmer harvests a field of grain, the nutrients end up in loaves of bread in a shopping center in some distant city. The forester and the farmer have to add fertilizer to the soil to make up for the nutrients that have been taken out of the system. We have already seen that some of this fertilizer is washed out of the soil and ends up in rivers and lakes, negatively affecting those food chains.

A more frightening prospect is the chance of radioactive elements entering the food chain. The 1986 accident of the Chernobyl nuclear plant in the Soviet Ukraine demonstrated how widespread and long lasting the effects of such an accident can be. The danger was most severe close to Chernobyl, but radioactive cesium worked its way into the food chain all over Europe, contaminating sheep, cows, milk, fish, and berries. One plant that was particularly affected was lichen that grows in the north in Norway, Sweden, and Finland. Lichen, or reindeer moss, is the staple food of the reindeer herds of the Laplanders. With no underground root system, lichens absorb their nutrients directly from the air. In late April and early May of 1986, they soaked up the radioactive rain like sponges. When reindeer ate the lichen, they concentrated cesium in their fat. The animals had to be destroyed, and their bodies buried, because their meat was unsafe. The long-term effect is still not known, but the immediate effect on the Laplanders has been devastating. Although they are self-sufficient and make few demands on the energy resources of the planet, they have not been able to escape the problems of the technological age.

How Many Mouths Can We Feed?

The number of people on earth keeps on growing. Every second of every minute of every day at least two more babies are born. Two more hungry mouths to feed. They are part of the food chain; part of the biogeochemical cycle of nutrients; takers from a limited pool of resources. There is no way we can keep on increasing the number of people in the world without adversely affecting the ecosystem.

In nature, the numbers of a species rise when its food is plentiful, but the availability of food or of niches eventually sets limits. Lemmings, the tiny grazers of the tundra, are subject to dramatic population increases and declines. In years when their population is high, migrant birds, such as the snowy owl and skua, stay and breed instead of moving on in search of other nesting sites. Arctic foxes and weasels eat well and also produce many young. The growing numbers of

Feeding the ever-growing world population depends on energy from the sun and from the fossil fuels used in today's intensive farming.

predators, as well as the fact that the grass and sedge have been over-grazed by the lemmings, lead to a crash in the lemming population, which in turn affects the number of predators that survive the following year.

We are more independent of fluctuations in food supply than are animals living in the wild, yet there are places where people die of hunger by the thousands when crops fail. We probably have the knowledge and ingenuity to provide enough food for everyone living in the world today so this need not happen, but the politics and logistics of distributing food are difficult problems to solve. Feeding the growing world population also requires energy—energy from the sun for photosynthesis and energy from the fossil fuels used in intensive farming and in the distribution of food to consumers.

The most serious problem posed by overpopulation is that we are all consumers—not just of plants and animals, but of the land itself. More people means bigger cities, more highways, more paved parking lots, longer runways—all of which leave less land for agriculture and less land for other living organisms.

Like it or not, we live on a planet with limited resources. When we take more than our share of nature's riches and niches, other species rapidly lose out.

And that means that ultimately we lose out, too.

Glossary

algae (singular, **alga**)—One- or many-celled plants without roots that live in fresh water or the sea; they can also be found on moist soil or trees.

anaerobic bacteria—Bacteria that live without oxygen.

bacteria—Tiny one-celled organisms; they can be beneficial, harmless, or deadly.

biodiversity—The variety of living forms occupying the same community.

biogeochemical cycle—A cycle that involves both living organisms and the earth, such as the water cycle.

canopy—Tree tops; the uppermost layer in a forest community.

carnivores—Animal-eaters or secondary consumers.

chemosynthesis—The process by which certain bacteria use inorganic chemical reactions rather than sunlight (photosynthesis) to produce energy.

decomposers—Organisms that get their energy from breaking down the wastes and dead bodies of other organisms into raw materials.

denitrifying bacteria—Bacteria that take oxygen from nitrates, and then release nitrogen into the atmosphere.

detritus—Dead plant and animal material, including feces and organic debris.

ecology—The study of the relationships of living organisms and their surroundings.

ecosystem—All of the living and nonliving parts of a given area in nature. Ecosystems range in size from puddles to oceans.

enzymes—Substances that can produce chemical changes in organic matter.

epiphytes—Plants that perch on other plants but take no nutrients from them.

feces—Waste material from digestion.

fossil fuel—Energy sources, such as coal or oil, formed from organic matter millions of years ago.

fungi (singular, **fungus**)—Non-green plants, such as molds or mushrooms, that take their nutrients from living or dead organisms.

galls—Growths on leaves or stems that are formed by insects or fungi.

herbivores—Plant-eaters or primary consumers.

guano—The droppings of sea birds, often used as a natural fertilizer.

krill—Small shrimplike crustaceans.

lichen—A plant that is composed of a fungus and alga growing together.

lignin—A tough substance found in cell walls of wood.

microbes—Microscopic organisms, especially disease-causing bacteria.

monoculture—Growth or cultivation of just one species, for example a wheat field.

niche—An organism's position and role within its community of plants and animals.

nitrogen-fixing bacteria—Bacteria that trap nitrogen from the air, making it available to plants.

omnivores—Consumers that eat both plant and animal material.

organic—Derived from plants or animals.

parasite—A plant or animal that lives on an organism or another species (the host) from which it derives sustenance or protection, without benefit to the host.

photic zone—The surface layer of the sea that receives enough light for photosynthesis.

photosynthesis—The process by which plants convert water and carbon dioxide into carbohydrates in the presence of sunlight.

phytoplankton—Small drifting plants in the sea or freshwater.

predators—Animals that eat other animals.

prey—Animals that are eaten by other animals.

protozoa—One-celled animals.

saprobes—Bacteria, yeasts, and fungi that decompose plant and animal tissue.

shredders—Animals that help in decomposition by chewing organic matter into smaller pieces.

trophic—Concerned with nutrition; trophic levels are the steps of the food chain.

zooplankton—Small drifting animals living in the ocean or in freshwater lakes.

Further Reading

How Animals Behave: A New Look at Wildlife. Washington, D.C.: National Geographic Society, 1984.

Lambert, David. *Grasslands.* Englewood Cliffs, N.J.: Silver Burdett, 1988.

Mattson, Robert A. *The Living Ocean.* Hillside, N.J.: Enslow Publishers, Inc., 1991.

McLaughlin, Molly. *Earthworms, Dirt & Rotten Leaves: An Exploration in Ecology.* New York: Macmillan, 1986.

Parker, Steve. *Pond & River.* New York: Knopf, 1988.

Reed, Willow, Ph.D. *Succession: From Field to Forest.* Hillside, N.J.: Enslow Publishers, Inc., 1991.

Schoonmaker, Peter K. *The Living Forest.* Hillside, N.J.: Enslow Publishers, Inc., 1990.

Smith, Howard E., Jr. *Small Worlds: Communities of Living Things.* New York: Charles Scribner's Sons, 1987.

Stevens, Lawrence. *Ecology Basics.* Englewood Cliffs, N.J.: Prentice-Hall, 1986.

Index